# ARTIFICIAL INTEL
# FOR PEOPLE IN A HURRY

# Disclaimer

# Table of Contents

# What is Artificial Intelligence?

## The Basic Concept

The basic concept of artificial intelligence is quite easy to understand because it's in the name. Artificial intelligence is intelligence that has been created by hand and not... artificial intelligence is intelligence that is not like real intelligence? Artificial intelligence is a machine attempting to emulate biological intelligence? Actually, the concept of artificial intelligence is quite difficult to nail down to a specific definition.

All right, the basic way that we understand artificial intelligence is an intelligence that is produced by a machine. This has been the standard definition for about a century. The reason why it is difficult to conceptualize a specific definition of artificial intelligence is that of how blurred the line is between biological intelligence and mechanical intelligence.

## The Application We Know

The most believable application that we know of when it comes to artificial intelligence is that of the enemies inside of video games. The problem is that this is not really intelligent. Or is it? The way that science defines intelligence is that it is an entity that is capable of attaining new information and then using that information with past information to define current information. Therefore, the primary problem with creating real intelligence is finding a way for machines to learn.

The reason why enemies inside of video games are still called artificial intelligence is the mechanical nature of enemies in video games. The average person who plays a video game will look at a first-person shooter enemy as being intelligent in that the enemy will follow them. Therefore, if the player begins to hide then the enemy will move closer while if the player is out in the open and shooting, the enemy will run for cover after firing a few bullets. This makes it seem like the enemy is actually thinking.

This is the purpose of artificial intelligence; faking the ability to think. The truth of the matter is that enemies inside of video games don't think at all. The easiest way that I can show you how a video game enemy performs these actions is with a little bit of pseudocode.

```
if(enemy.line_trace_end !=
player.collision){
    enemy.position moves closer to
player
}else if(player.line_trace_end ==
enemy.collision){
    enemy.hide * ((Random.new() * 10) +
1)
    if(enemy.hide >= 75){
        enemy.position moves behind
nearest collision
    }}
```

As you can see from the pseudocode above, if the line trace of where the enemy is pointing to does not equate to the player's collision but another object collision, then the enemy will move forward. This brings the enemy closer and creates the illusion that the enemy knows the player is hiding. However, if the player is targeting them and the line trace meets up with the player collision box then the enemy will decide whether to fire at the player or hide behind a collision based on a randomized value. The line trace is the path of motion of the player.

The randomization creates the illusion that the enemy chooses whether they are going to duck underneath cover or they're going to fire at the enemy. However, this is fake intelligence if we consider the fact that there is no past information being involved here. This is current information dealing with current decision-making. Additionally, the machine doesn't actually have any ability to learn about the character that they are going up against. This means that while the machine looks to be intelligent, there is no intelligence at play. Instead, you have an instruction set pre-programmed by a thinking individual that tells the machine what to do given a certain case or a certain circumstance.

Now you might think that something like Amazon Echo or Google Home are hubs of real mechanical intelligence. The problem is that these machines are working very similar way to the enemies that we just covered.

Human speech is dictated by rules, you can think of them as the instruction set for how to speak the language. The important thing about rules is that if there are rules, a machine is most likely going to be able to follow those rules. Well, I won't get into the specifics of how recurrent neural networks work, but you can basically assume that the machine is simply analyzing and making probability statistics on what words likely fit together based on those rules. This is a very accurate way of determining which words you, as a person, are going to say whenever you use something like voice typing or when captions are made inside of a video that have been auto translated your words.

Up until this point, we have not had many machines that are capable of learning and using that learned knowledge to further improve. However, we have just now entered the digital age of utilizing something called backpropagation.

Backpropagation is the center point of nearly all machine learning algorithms that are currently at play. A neural network is designed to take database input and perform actions on that information inside of neurons that then produce an output. If the output is wrong, we then change what is occurring inside of the neurons so that we try to get to a more optimized answer. The standard way of doing this is by taking the variables inside of the neural network and creating sets of randomized variables to find the best-randomized variables. This is a very slow process and usually takes a very long time in order to perform. The new method is called backpropagation, what I just described to you is called forward propagation.

Instead of utilizing randomized variables and never knowing if you have reached the ultimate randomize combination, we use calculus and the outputs that the neural network has provided. The information is fed backward into the neural network after the answer is performed incorrectly and the neural network takes the incorrect results as well as the numbers they used to get to those results and creates a combination that is more optimized as a result. Backpropagation represents the first form of machine learning that is actually intelligent. However, the vast majority of neural network applications, like Alexa and Siri, are not normally back propagating networks in most situations. As time progresses, this actually might be changed but backpropagation takes a lot of computing power that most devices simply don't have.

**The Applications We're Creating**

I've talked about the different types of artificial intelligence; emulated intelligence, probability intelligence, and machine learning. A lot of individuals tend to categorize all of it as artificial intelligence, which is incorrect. If you remember correctly, artificial intelligence is where an entity attempts to emulate the ability to learn without actually learning. Machine learning is clumped in with artificial intelligence, even though there are certain forms of machine learning that actually don't fit in artificial intelligence. You have several different robot projects that are based in machine learning that are no longer artificial intelligence.

However, we are not actually here to discuss that realm of intelligence. Instead, we are talking about applications where artificial intelligence is being used. You have various different industries that are currently massively affected and will be massively affected for the next decade as new artificial intelligence mechanisms come out to affect that industry. You have a self-driving car, robotics that are designed to emulate companionship, and you have loads more that we will talk about in the next chapter. However, we do need to talk about the pinnacle of all artificial intelligence. Artificial intelligence is not possible without information.

When I gave you the example of the pseudocode, you will notice that the choices the enemy actor made were based on information about the current situation. This is true of all artificial intelligence machines because nothing can be decided if nothing is coming in. We, as humans, receive input all the time from the world and artificial intelligence machines need that same input and sometimes even more.

## The Information Age

Knowing that artificial intelligence needs information and given the current political climate that this book is being written in, we can conclude that we are in a new era. You have Facebook trying to collect as much data as possible while Europe is trying to stop Facebook from collecting so much information without the permission of the user. These internet-based companies have had unfettered access to the information of our daily lives because it was simply assumed that the user would understand what they were giving up here.

The problem is that the average person is lazy and ignorant most of the time. A person may find that insulting, but that is how the market works. A person buys a washing machine because that's what their parents did, or they don't want to wash their clothes by hand. A person buys a dishwasher because that's what their parents did, or they don't want to wash dishes by hand. There was a man that I conversed with once that did not know where beef inside of a grocery store came from, he simply assumed the grocery store made the beef. I'm not talking about where he didn't know which company shipped the beef to them, I'm talking about he did not know that beef came from cows. This was a person who worked in the government and yet he did not know this very simple, commonly understood fact. There are many people like this. Those individuals who have a lack of what would be normally perceived as commonly understood facts purchasing the items these technology companies make. Markets rely on ignorance and laziness to work. After all, why hire a mechanic if you know how to do it yourself?

People tend to skip the end-user license agreement because it is a huge document that is not very entertaining to read. There is a very small portion of the crowd that will read that document and that portion of the crowd is generally how companies are kept in check. That person will read it, find something horrendously wrong with it, and then share that information with people who don't read it. The general market relies on people not being able to fully understand how things are done. If you bought a t-shirt that was $100 and thought you were getting a

19

great deal, you might find you were instead getting a horrible deal once you know how it was made. You can buy a standard t-shirt off the internet for $5. You can then go take a picture you want, and have it placed on a special paper that you can then heat press onto the shirt for about $5. That means, practically any shirt that you buy that is simply a logo without being any special type of gel and is just a plain image is usually about $10 in cost. Now, there are additional costs, like shipping and branding, but it took $10 to make $100. If you know how to do it yourself, you could have saved an additional $90.

However, you are likely not going to go out and make every t-shirt that you want to own unless you are incredibly stubborn and self-motivated. Most people will simply claim they don't have the time to do it. However, the fact of the matter is that that's a lot of money that is being thrown away due to laziness as many of those same individuals will binge watch a 12-hour season. Restaurants rely on people not making recipes, clothing stores rely on people not making clothes, and tech companies rely on people not creating tech. This is how the market works and how it has worked forever. However, the new commodity is information as information is now how most companies make their money.

## The Fourth Industrial Revolution

Facebook sells your information to advertisers so that advertisers can better promote the products they are trying to sell. Google actually does the exact same thing but has different data. Almost every platform does the same thing, except that most of it has been allocated to a handful of companies. The standard website will not go out of their way to create an advertising tree that advertisers can participate. Why would advertisers go to an unnamed website to promote their ads? No, most standard websites that are trying to make it online will utilize Google ads or Facebook ads. This allows the website to gain money from Google or Facebook by selling advertisements.

Another section of the internet is affiliate links, where people have audiences that they can direct traffic to specific products. The most notable of affiliate links are the Amazon affiliate links, which seem to be practically everywhere. Internet Stars are able to sell products of other people (like an advertisement would) to crowds that would generally be interested in it. This is much more effective than simply trying to blast on the internet that you have a specific product that you want to sell. In this situation, the information is simply that you know what your crowd likes to see and what your crowd is after.

We can't get anywhere without Alexa or Google Voice or Siri because most of our information goes through these three individuals, machines, to produce what we want. We ask Google pertinent questions about our daily lives. They collect information based on our search results and our requests so as to better serve advertisements to us. However, they are not the only customers of information.

You have subscription services like Netflix and Hulu that utilize your watched patterns to determine which videos you are more likely to want to watch. By serving you what you want to watch, these companies keep you on their platform for a longer period of time. In fact, it's so bad that there have been cases where people who watch Netflix go into rehab to overcome their addiction to the service.

Information runs practically all of our businesses as of right now and the person who can collect the most information is often the biggest business on the internet as well as in the real world. If you look at Google, Google is supposed to be a search engine, singular, that provides advertisement on the sides. Yet, they are a massive company being sued for monopoly reasons. If we rewind history back to the 1960s, Google didn't really exist, and you had cable companies and telephone companies being the biggest alongside real estate and other industries. The tech industry, which solely relies on information to actually work, is the biggest industry in the entire world. It's in everybody's home, everybody's kitchen, everybody's room, everybody's car, everybody's road, and I think you get the point.

# Our Daily Lives with Artificial Intelligence

## Finance Industry

The finance industry is perhaps the most well-known industry for utilizing artificial intelligence before any other industry decided to use it besides the computer industry. It was long thought that stock markets and pricing graphs could generally not be predicted, but a serious investment into prediction algorithms has had sway over the highest companies playing in the stock market.

In fact, it's become so common knowledge that these companies are vying for prediction algorithms that other people, as individuals, are also trying to get their own prediction algorithms. This is because the prediction algorithm can actually be taught to be more accurate than the individual.

However, that's not all the artificial intelligence community has dabbled in with the finance industry because there's quite a bit more. A little bit closer to home is monitoring the habits of the individuals who hold money within a bank. The number one concern when dealing with bank accounts is whether there has been fraudulent activity on those accounts. Due to the fact that banks are often the first financial industries to get hit, besides credit cards, understanding habits that are out of the norm are a key aspect of determining whether a transaction was fraudulent or not. With artificial intelligence, the machine can be trained to understand what is most likely to happen in the case of that specific person having a specific amount of money. For instance, if you have a lot of money at once, you might be the type of person that is a little bit conservative with their money or you might be the person that goes to the mall and blows it all. If you are more conservative on most occasions, the artificial intelligence will get really suspicious if you decide to go blow it all.

The idea of artificial intelligence in the stock market is actually just a segment of what A.I. does in what's known as advisory artificial intelligence. It's a much bigger category and it includes things such as marketing forecasts, audience expansions, and a lot of other business technical terms. The purpose of an advisory artificial intelligence is to train the artificial intelligence to predict what has not happened based on what has already happened. The crux of this type of A.I. is that almost all situations dealing with predicting what has not happened is based on context and so it is difficult to quantify context, but there are several applications of artificial intelligence in advisory roles.

Perhaps the less talked about part of artificial intelligence in finances is the A.I. that helps in finding suspicious patterns with companies. This isn't really talked about because it's not really something that the public is concerned with but rather governmental agencies looking into the public. Previously, the way that white-collar crime could be detected on its own is by looking at the finances of a company and seeing where the money is going. This is a very laborious and man-hour eating task. Artificial intelligence is so efficient and quick that what it could take a hundred-man team to do in a week could be done by the artificial intelligence in maybe a day at the longest. Therefore, A.I. not only is helping out the public with their finances but also helping the legal team in catching criminal activity in companies through their finances.

## Sex Industry

Artificial intelligence has made leaps and bounds within the sex industry. If you rewound time to about a decade before this book was written, you would be lucky to meet a chat system artificial intelligence that you could pay by the hour to turn you on. As time has progressed, there have been several sections within the sex industry that have expanded the purpose of artificial intelligence in ways that you would expect and ways that are unexpected.

Of the ways that you would expect, these chat system AIs have gotten better at forming conversations with people. There is a particular type of technology called MMD that when combined with artificial intelligence can simulate a FaceTime interaction. In fact, the aspect of MMD has expanded into another software called Facerig that I see expanding even further into a new type of genre that has recently opened up in Japan. In a product known as the Gatebox, a virtual girl or woman can do practically anything you would want them to do as a virtual companion. You can buy your favorite virtual women or men to put inside of this new technology so that you can come home and interact with them. This is referred to as companionship AI, which is slightly different than assistant AI.

If you had such a technology, you could expect to come home and be greeted with a warm and friendly hello or welcome home from your virtual wife or husband of your preference. This technology would act similarly to Amazon Echo or Google Home in that it would remind you of certain activities or dates, but it would also interact with you on a more personal level. For instance, it would send you text messages saying that it looked forward to you getting home or asking you how you are doing in the middle of your day. Similar to how relationships might work, this artificial intelligence is specifically designed to make you feel better throughout the day.

Carrying this emotional artificial intelligence over into the world of Life-Size dolls otherwise known as sex dolls, we can see that there are several developers looking to imbue sex dolls with emotions. To somebody who hasn't looked at the industry since a documentary was made about it on television, the industry itself has primarily consisted of people who invest money in sex dolls because they prefer a relationship with a doll to be more preferable than a human. One can understand that kind of logic because there are some drawbacks when dealing with a human. A lot of people don't really like interacting with individuals who provide them with negative attitudes and negative opinions, which naturally leads to seeking comfort in other things than the people around them

In the emotional AI of the sex dolls, the human that owns said sex doll can not only interact with the sexual part of the doll but also interact with the emotional elements of the doll. For instance, the doll could be sultry but also caring or it could be a complete tsundere with a little bit of light humor. Essentially, you interact with this sex doll and its personality conforms to What would most likely be best for your personality. Now, there is an option for you to have customization over it for sexual purposes, but if you're looking at it for a companionship role then you can have them adapt to your own personality.

Finally, the most interesting part of this is that there are now devices that allow you to have sex with computers. Now you might think that the previous section that talks about sex dolls referred to having sex with computers, but I am talking about devices that are hooked up to an artificial intelligence to specifically pleasure you. For instance, a recent development of a product for the virtual reality world was called into question in a very weird light. Essentially, the girlfriend of this man had accused the man of cheating because the man used a device meant for masturbation in the virtual reality environment. Imagine a woman being recorded by a man (in the first-person view) masturbating that man and a condom with sensors on it is recording the pressure of the woman's hand. All of this is recorded and when a paying customer purchases this, they can feel the same sensation of that woman's hand on their crotch only it's through a device instead of a human hand. You might be wondering where AI comes into this. Well, there are small segments of the internet where this is not a recording but an MMD character that's using this same device to learn how to properly masturbate the client. That's right, a virtual reality AI can now be used in substitution for, basically, digital prostitution.

## Health Care Industry

The healthcare industry benefits significantly from artificial intelligence as artificial intelligence has had significant improvements in image recognition technology. In our everyday lives, image recognition technology provides us with little tags that we can put alongside the faces of the people that we take pictures of. In medicine, the story is a little different as doctors tend to take pictures of bodily organs and parts that might have something wrong with it.

The most common thing that the health industry is using artificial intelligence for is to recognize types of malignant cancer better than human doctors can. The benefit of this is, of course, saving lives, but the truth of the matter is that artificial intelligence does not need to have you be in the same room in order to make a diagnosis. To individuals who do not regularly go to the doctor, you will likely not know why this is important, but this is the reason why Doctor Services are not normally found online. Doctors insist that they get a physical examination from you, which means they often want you in the same space as themselves to give a proper diagnosis.

While it is currently only being used to significantly improve the ability to detect malignant cancers, this technology is going to eventually come down the health industry ladder to the average individual. It is commonly known that an individual simply needs to look into the mirror with their mouth open to see if there are white spots on the back of their throat to give an estimated answer as to whether they have strep throat. Now, doctors often also insist on a saliva swab to ensure that the diagnosis is correct, but many diagnoses can be given in certain circumstances that only require a visual. For instance, a common cold is usually looking at a patient and taking their vitals. However, going to the doctor for such a visit is usually way more expensive than people can justify it. After all, if a doctor visit is 1/3 of your weekly paycheck, you are not likely to go if you make below a certain amount of minimum income.

The problem with this is that people tend to go to work sick even though they should technically be at home. Common experiences from fast food chains has shown that a good portion of the workforce working at minimum wage simply cannot afford to take the time off. Now, obviously, for the sake of not spreading their illness to the food that they are making they should stay home but this is not normally seen as the more important of the two concerns in such a situation by the individual's perspective. To the individual impacted by the minimum wage that is. However, if an individual could go to an online service that was relatively cheap per visit or even just had a monthly subscription that was reasonable, they would probably get seen by an artificial intelligence doctor so that they could get the medicine to make them feel better.

**Transportation Industry**

Recently, we have actually seen big industries push for automated motors that no one except a computer has to drive. Now, there are a few industries that I want to talk about because the transportation industry is a huge industry, consuming most of the market we know of today. You have consumer cars, the technology going into these cars being equivalent to technology going into a tablet, virtual assistants being inside of cars, and the list really does go on for the automotive industry.

The first one that I want to talk about is the trucking industry. Recently, I got to experience a surface level simulation of what it's like to be a truck driver via the game *Truck Driver Simulator*. I finally understand how difficult it is to drive a semi from the origin to perhaps hundreds if not thousands of miles to the next destination. It involves balancing tons of items that precariously hang outside the back end of a semi. The easiest part of the job is hooking the trailer up to the actual semi. The hardest part of the job is a mixture of corners and dropping the material off at where it needed to be. Psychologically, the hardest part of the job is simply being on the road because some of us travel hundreds of miles every few years visiting different places, so we get a new experience. However, people in these positions often have isolation problems because they can be alone for around a day or two at the minimum.

This does not stop the potential automation for these jobs. This is a decent paying job, usually paying around $20 to $30 an hour as a minimum rate in the United States of America. It's not an easy job, even though it may look easy. As I mentioned before, the hardest part is from the psychologically of driving all of those miles to get to your destination. This job may be automated, in the near future, around this central part; driving to the destination. Elon Musk and Uber have announced they plan to make self-driving a part of their future technology. However, these are the companies that practically everyone knows about. There are several other companies fighting in this self-driving game. Specifically, for semi driving, you have companies who employ these people trying to automate the service.

Now, you have also got to realize that semi driving is extremely dangerous for both the driver and the people around the driver. This is why semi drivers have to go test regularly and generally have one of the hardest driving tests of all the vehicles that there are. However, that doesn't stop these companies from trying to automate an industry that has previously laid untouched by the newest self-driving algorithms.

With self-driving semi's, the overall cost of food items shrinks dramatically. A semi would likely be carrying several items, but let's say that it's 50 items. If a semi driver has to drive for 3 days or 48 hours, then

$$48*30 \; / \; 50 = 28.80$$

is added to the products in that semi (for labor only). Companies don't normally add $28 to a stick of butter, but they will add $46 to a TV. As a semi is the main form of transportation for almost all in-store goods, that cost goes on all products. This means a $1,000 television could be $800 or a bag of cheese goes from $10 to $5.

As I previously mentioned, Uber is in that category as one of the companies that are seriously investing in self-driving cars. As a customer of Uber, one can see that the benefit would be on both the customer and the business side, but not the worker side. Many of the individuals that Uber are either doing so because it is the best option for them or because they're bored in retirement. On the customer side, the ride will get cheaper because you no longer have to pay a driver and on the business side it will get cheaper for the same reason. However, on the business side, you can also profit by being able to have many more cars than you have employed people. The public will almost always want to have a person driving them because of the technophobia and the potential personal conversations one might have while they are being driven somewhere. It's actually something that you kind of pay for that companies don't even realize.

However, if you also recall, I talked about how dangerous automation of driving truly is. Uber has been under some scrutiny because one of their drivers wasn't doing their job properly and the self-driving tester car managed to run someone over (that's how I understand it anyway). This is also there for semi driving because you have an automated vehicle on the road. The only difference is the extremity of how dangerous this is. It is much more likely for a semi-truck to cause an entire roadway of damage than it is a single car. However, as we push forward as a society, it is much more likely to have vehicles with emergency mechanisms so that you can take over should the vehicle not drive properly.

## Education

## Focus Areas

One of the key problems with education is where the student is actually placing their focus. For instance, I knew a student once that really loved reading but often loathed doing any sort of math. The instructor themselves didn't concern themselves with how this student was failing in this selected area of the study content the instructor was giving. Once I had the student explain to me what he did throughout the day and then also explain to me the feelings he had about the mathematical part of the content, the reason to me as to why he was failing in that part of the curriculum became obvious. He loved reading because he enjoyed going into a world that allowed him to escape, but mathematics is often taught from a listen and do perspective.

Once I suggested that the student not rely on the instructor and instead read the material rather than listen to the instructor, his mathematical scores significantly increased. The reason for this failure was simply due to how it was delivered. I can foresee an artificial intelligence system that takes over consultancy as we have actually seen in artificial intelligence machines built to diagnose patients. This field of artificial intelligence consultancy is relatively new and still experimental, but I can find something as simple as what I found then an artificial machine, with the same variables, will undoubtedly be able to draw a similar conclusion. However, the conclusion will be brought about in a significantly different way until machines can understand Context. Context is the last field of abstraction left in the artificial intelligence pipeline for improvement.

**Teaching without Teachers**

This leads us into the other area in which artificial intelligence is both currently affecting and will affect teachers in the future. As of right now, artificial intelligence is being used to predict which areas of statistical analysis are either the weakest or greatest. Therefore, if a lot of students seem to be failing at a very specific part of your mathematical class, say algebraic functions, this would only be noticeable to a machine that could calculate the number of students multiplied by each test answer. Now, that may seem like a very simple Equation to perform, but the equation is much more complex than that. You would need an algorithm that can quantify each answer as both parts of a group and an individual, which would lead into determining whether a group of errors is a common pattern or a group of independent errors is a pattern. Such a distinction is easily made between two students.

For instance, if a student of Class A gets functions wrong and students of class B manages to get one function wrong, we would naturally conclude that students of Class A is having problems with functions, but students of Class B simply got the answer wrong. By having artificial intelligence be able to look at this, artificial intelligence could then create a course plan for students individually based on the topics the artificial intelligence is teaching. The purpose of a human being in a teaching position instead of mandatory reading is so that there is a guide for the person that is learning and a person to set the goals of the class. An instructor is a person who simply tells the class what they need to do, a teacher is a position held by those who act as instructors and those who guide students through those instructions. However, having an artificial intelligence that could figure out whether the class as a whole is having a problem, an individual is having a problem, or a particular topic is doing rather well you could have that same artificial intelligence change its own curriculum plan to better teach students. The artificial intelligence can already act as an instructor by default, as many help prompts have told us in the past, but by being able to optimize a curriculum the artificial intelligence also becomes something that can guide those instructions.

**Journalism 425**

**Aggregate Information**

One of the main purposes of Journalism is to aggregate information into a central location and generally provide it objectively. Notice that I said generally because in the current journalism environment, this is rarely done. It is rare to see journalists simply presenting the facts and then distinguishing between the facts and their opinions. The best example of aggregate information being carried out by artificial intelligence is the Google you use every day or most of you. Google does something called sending out a web crawler, which is a fancy way of saying that an advanced algorithm designed to find websites on the internet and include them inside their search listings. Over time, they have created algorithms to determine the accuracy and importance of them, but the primary example here is to show that artificial intelligence has been aggregating information in the past, but this is known as dumb artificial intelligence. New artificial intelligence will be able to collect specific information, most notably the information related to news. It will then be able to follow leads of that news on the internet to see if it's related to any sort of data. For instance, statistics on crimes or number of bakeries in a city on average. This leads us into the current climate of Journalism as we know it in the political sphere.

## Accuracy Judgement

There has been a massive push to censor what's known as fake news, but this has a few problems attached to it. The big problem that everyone practically knows about is how to determine what is fake and what is not fake. The obvious answer that most people reach for is that if it is true, more people will be talking about it. However, you have flat earth believers and round earth believers that have constant debates about Earth. If you were to go and search the amount of Flat Earth information, there are more websites that consider Flat Earth to be a reality and there are practically none about round earth because those who believe it is round don't feel the need to convince people. The round earth believers simply assume that anyone with a higher IQ would reasonably believe this and just accept it as fact. Therefore, in the online space, you have more people talking about Flat Earth being a fact than you do round Earth people. Following this obvious law, you will quite literally censor the people who believe in round Earth. Now if you felt one way or the other about the information I just provided beyond completely static, you will understand why it is important not to censor ideas. This is actually the reason why the free speech in America is so protected because the public generally wants what they consider to be crazy out where they can fight the idea rather than have that idea in a position of power and not be able to stop it. There are artificial intelligence systems being created to determine fake news, but the specifics of those facets of artificial intelligence haven't really come to light just yet.

**Automated Social Media**

While the specifics of accurately judging journalism aren't known, artificial intelligence can easily take over automated jobs. For instance, when sharing an article amongst the many social media websites, most websites simply have a single button attached to the publish button that allows them to just instantly post to all of the websites at the same time. This makes the whole process easier and thus saves the need for controlling whether websites get published on social media on a regular basis.

However, what has not yet been achieved but has mostly been theorized is automation of social media activity on those social media websites. The difference between sharing a post that was written and being active on a social media site is the difference between writing a story and writing a tweet or a Facebook post or whatever website you use to update everyone in your life. Artificial intelligence is currently at a crux in existence because we don't quite yet know how to define the context. That does not mean that a social media account wouldn't be able to regularly post tweets given a set of rules.

For instance, one of the common rules that I was taught during low-level grades is to take the question being asked and reword it as my first sentence. This is a rule that can easily be followed by an artificial intelligence as we have many reword websites that allow you to take sentences and reword them. If someone posts on social media, an artificial intelligence algorithm would be able to take what they said and reword it according to rules. Additionally, they could post things that your website might talk about in the next day based on titles that you wrote today. Essentially, the artificial intelligence would be generating leads that would entice users to follow that account because they're interested in what you said. Social media for journalism is both a medium for spreading information but it's also a medium for advertising the journalism that the journalists are doing. This is already being done today in marketing campaigns by activist groups to hide how small that group might be. We heard a lot of rhetoric in the presidential election of Donald Trump about social media bots doing just that.

## Agriculture

### Better Forecast Predictions

One of the most wasteful things that a farmer has to deal with is just how much they need to plant as well as just how much they need to harvest as well as when they harvest. In the United States of America, a lot of the food waste goes to food that is neither appealing or used. In fact, this is a common problem in developed nations in the world because most of the developing nations work on a monetary system. If you harvest too much, you will end up wasting what would essentially be the monetary value when you can't sell it. If you don't plan enough, you might lose out on potential gains you could get had you grown enough.

This is an equal problem for the people eating that food because even though you only need to produce a certain amount of food so that the market sustains itself, there are people in developed nations still

dying from starvation. It is an odd problem when you have so much

food that you are throwing it away and yet you continue to allow

starving people to die. The problem is that the market isn't entirely

predictable and so you will have waste, or you will have shortages, but

you will never break even. Artificial intelligence may not be able to

break even, but it can provide farmers and those in the agriculture

business with better predictions on just how much they need to produce.

The easiest way that many of them use is simply to grow as much as

they can and then use what they can't sell. This works out for a lot of

small farmers, farmers with around 100 acres to 500 Acres. In these

small localized areas, they are still able to use most of what they grow

up and make a decent living off of what they sell. The problem really

comes into play when you go above 500 Acres. A 4-person family is

not going to be able to consume 900 acres of corn in the next 6 months

unless they eat nothing but corn for the next 6 months, but if they sell

another type of crop it is virtually impossible. Also, that family will

probably die because corn doesn't have much of a nutritional value. In these circumstances, it would be better to figure out a trend line for what they need to plant and how much they need to harvest as well as when to harvest. While farmers do a fantastic job at providing food for a nation, not a lot of them are statisticians by trade. In fact, because of how much professional statisticians make in various roles, it would be more lucrative to be a statistician if you were just out for the money. Therefore, the best alternative is to provide a product that will make those predictions for you and this is where artificial intelligence comes up. Artificial intelligence already has a huge involvement in the stock market, so utilizing it for something a little bit more predictable like farming has been an easy crossover.

**Advanced Growing Techniques**

Farming doesn't really have many scientific advancements as to how food should be grown but if the farming is done in a controlled environment, these variables can be handed off to an artificial intelligence machine. The reason why such a fact is important is due to how artificial intelligence machines are designed to optimize whatever they're handed. A lot of farmers work is actually based on guesstimation or estimated guessing. For instance, many farmers will look for a specific temperature that holds consistency to plant items that are temperature sensitive. You can only grow potatoes during certain seasons based on your location. An artificial intelligence machine could draw on forecast weather and predict weather patterns, which would lead to faster planting of the productive plant.

Because artificial intelligence can be handed Sensors, they can also handle when the soil needs to be given a specific type of nutrient. Normally, farmers rely on the look and condition of a plant to determine when that plant needs more of a specific element. Usually, the plant will show signs of yellowing and wilting. However, if you have sensors in the soil and you have visual open recognition monitors looking at plants, you can teach artificial intelligence to look out for the same things. A crop may actually go for a couple of weeks without the proper nutrition because the farmer simply doesn't notice the signs that quickly, but an artificial intelligence machine would be able to know almost immediately that something has gone wrong and what has gone wrong on the day that it begins. This would allow Farmers to produce more product and reduce the amount of waste, which would generally benefit everyone. This is why, within the past decade, this new form of artificial intelligence has taken root in the more tech-savvy farmers in the market.

**Enhanced GMOs**

The last bit in which artificial intelligence can help is when developing new genetically modified organisms to better grow for their environment. There's been a lot of stigma around GMO's, but the truth of the matter is that a lot of what the world eats today is made up of GMOs. For instance, most corn could not be sold on the scale that it is without being a GMO of the original corn plant and a weed. In fact, some consider corn not to be a vegetable but a weed because of its origin of being mixed with a weed.

We partake in genetically modified organisms everywhere even though it's considered bad by some. Our children are genetically modified because they are neither all of us or the significant other required to make one. Our food has been optimized to grow more and grow bigger by mixing it with breeds that last longer, provide more food, and provide bigger food. It wasn't until recently that we began using a lab to create GMO products, but we have been doing GMO practices for centuries at this point.

Knowing this and knowing that it is a science, we can build artificial intelligence machines that predict better combinations for the future. This leads to more food and thus more product, which also means that we can sustain more population as a result. This is actually already being done, it's just not widespread yet and a few massive corporations have dedicated their resources to perfecting this.

## Law

### Better Defenses

There is currently an artificial intelligence known as Lisa, which acts as a robotic lawyer for creating legally binding contracts. Due to artificial intelligence being able to access a wealth of information, they are able to formally design rules based off of past rules. It's actually not that difficult for an artificial intelligence to create a binding contract because of a few elements.

The first element is that many contracts are almost identical, and you can see this when you deal with websites that have autocomplete forms with the contract. The second element is understanding the parties involved and the legal relation of the importance in a document. The last element is because, ultimately, most parties will proofread any document provided by a robot lawyer.

The first part is that many legal documents are identical, and you can actually see this commonality when it comes to software. There have been a few notable software companies that have just stolen the contracts developed by bigger companies as the way to save on money. They need, generally, the same things because they are a software business and so most of the elements affecting other software in legal terms should technically affect their company. Due to the standardization of legal documents, it is easy to figure out what contract you need. This choice can also be easily made by AI based on questions, such as "What type of company are you?" or "What are you attempting to achieve with this contract?".

The second part is that the relationships between parties are generally easy to understand. A software usually has one type of relationship to a user just as a real estate manager has one type of contract with the person that hired them. There are more complex relationships, but for most of everybody, there's usually a one-to-one relationship when it comes to a contract. This person wants something from this other person and needs a contract for it. Due to the simplicity, Lisa is currently able to provide non-disclosure agreements and property contracts as a result of this.

The final element is that we simply don't trust robots completely yet because we've been given plenty of the examples not to do so. Therefore, all contracts made by an artificial intelligence will be proofread by the person who asked for it and then the person who asked for it will provide feedback. This allows the company that develops this artificial intelligence to further optimize their artificial intelligence to better fit the needs of their customers. Therefore, as customers come in the artificial intelligence not only gathers data from those customers but also gathers data from customers with complaints.

## Better Rule Enforcement

It is generally seen as bad practice to be a bad lawyer for criminals, but it also is seen as a bad practice to be a good lawyer for criminals. You might think that this sentence is conflicting, but I am talking about two different perspectives. To other lawyers, it is a bad idea to be a bad lawyer for criminals because then you are just a lawyer bad at their job. To the public that doesn't like the criminal, you are a lawyer defending a criminal and so you are often seen as bad as the criminal and sometimes even worse.

Robots don't have to care about this perspective issue because they are robots. This allows them to truly be impartial when putting together a case to be used in court. There is an application known as the *DoNotPay* application, currently found on the Apple Store, that allows you to get free legal advice. There is a legal firm known as ROSS

Intelligence that allows legal teams to analyze documents for legal processing much faster than they would be by a team of humans. These are new technologies that are currently experimental but have shown not only can they do it faster and better than most humans, they can also do it for practically free. That isn't to say that companies won't try to make money off of these technologies, but most of what I just introduced is actually charged to the customer. Normally, a law firm will charge after the first hour for consulting. A law firm using humans to review documents will usually bill a customer for that. By having robots do this, not only do the law firms gain unbiased reviews to the information but customers also don't have to pay exorbitantly high amounts of money to hire a lawyer. A small firm could use these technologies to help their local city have effective defenses to potentially larger companies or people.

# Business Processes with Artificial Intelligence

## Smart Virtual Assistants

We have had virtual assistants for almost a decade but almost everyone agrees that virtual assistants aren't very useful. The truth of the matter is that they are not really built to be that useful. However, it depends on what you want out of a virtual assistant as to how useful they can be. You see, you can utilize virtual assistants to schedule meetings, read emails, and generally do a lot of office related items with.

The reason why most people don't find virtual assistants very useful is that they often just utilize it to answer questions they might have. While that is definitely useful, a virtual assistant can take notes on business meetings, act as a recorder for sessions, send mass emails to everyone on the business account, and do much more just as a real assistant would. The only problem is no one really knows the best way to talk to a virtual assistant. Most virtual assistants are activated by speech, which is a contentious issue whenever you have any type of accented language involved. The way that these virtual assistants are trained is by having a bunch of voice actors come in and use their voice to match up words to sound patterns. You can actually train a virtual assistant to recognize your accent, but it takes a lot of work. However, you will find them generally used for doing any office related activity that you might have in your business. You can even have your virtual assistant read a book to you if you really wanted to, it really just depends on what you want out of it and most people find virtual assistants to be almost as usual as a human assistant. The only thing that I can think of that a virtual assistant can't do for you just yet is write an email.

**Market Research**

A huge portion of machine learning, and artificial intelligence is specifically designed to automate market research and refine it. Most market research done by artificial intelligence, provided that it has been trained correctly, is more accurate than the Market analyst can be at making the same prediction. This is because the computer analyzes more numbers faster than the analyst can and the computer doesn't make mathematical mistakes. Normally, when you ask a company to give you a market report they will come back in about a week with a long piece of literature to explain the market for you. However, if you utilize an artificial intelligence to do the same thing, you will often find that the report takes a few minutes to generate provided you have the right data.

Additionally, thanks to algorithms like the K cluster algorithm, you can find General Trends and successes where it would normally seem like nothing was going on. This is often used to identify successful companies in areas of interest that would normally go unnoticed.

**Chatbots**

Perhaps the best source of artificial intelligence inside of the business environment comes in the form of chatbots. If you want to cut down on the amount of Human Resources allocated to customer support, then a chatbot is usually the way you want to go. Chatbots that are artificially intelligent and not just pre-programmed lines of script

have a wonderful way of convincing other humans that they are talking

to a human. Most notably, the response time from a chatbot versus a

human is significantly lower and thus many more people come out of

talking with a chatbot feeling a lot better about the company.

Having said that, chatbots usually aren't that intelligent. If you

are going to hand chatbot something to do, you should probably hand it

the most common items that your company gets called for. The reason

why you want to keep it simplistic and why you want to find a way to

remove the most common items from the call list is because chatbots

are not intelligent but will reduce the amount of calls your call center

will get.

**Email Autoresponders**

As I mentioned earlier, virtual assistants are currently known for

being able to provide the basic necessity of sending out emails.

However, you likely thought of those remailers that simply sent out

emails responding to password or email resets. The autoresponders currently coming out are a little bit more advanced than that. You have autoresponders capable of opening help desk tickets, answering common questions like software pricing, and even ones that will inform clients of available times they can make appointments. Some are so good that an actual person rarely needs to be in contact with a customer, which saves countless hours on the customer service end.

While I say this, we also know how horribly wrong such a service can be. In truth, this is a common occurrence as artificial intelligence can't rethink an existing solution. Therefore, when an oddity occurs the artificial intelligence cannot self-correct, which can increase the dissatisfaction a customer can experience. This is often remedied by ensuring that the customer only interacts with the autoresponder for a given number of messages.

**Taxes**

Taxes are the bane of everyone except the oddities that find it fun, but everyone finds something fun, right? Artificial intelligence exists based on rules, which means that a tax system, an exorbitant system of rules, is kind of like the bread and butter of A.I. However, there is a slight crux to Tax A.I. Due to the human nature of constantly changing practically everything, taxes change on a yearly basis, which has to be accounted for by the software. This has been what has taken software companies so long to create an efficient A.I. that's capable of doing an individual's taxes. Therefore, this is one of the new frontiers for A.I. as companies have begun rolling out new software to fit this need.

However, depending on how it is deployed and implemented, a company that uses the software may not even need to fill out paperwork. Since you have to give the A.I. your input, it could be capable of doing all the taxes for you eventually. However, most tax A.I. is centered around determining how taxes will affect the business of the one forecasting the prediction and/or the economy that individual/company is in.

# Self-Driving Cars

## What are Autonomous Vehicles?

### What it means to be Autonomous

To break down with a self-driving car is, we first have to actually understand what it means to be autonomous. The actual word itself is comprised of two Greek phrases. The first phrase, *auto*, means self. The second phrase, *nomous*, means customary or law. Therefore, one can naturally conclude that the phrase literally means self-law or self-custom. By itself, it doesn't have much meaning. It is only when applied to an entity that is capable of thinking that autonomy actually has a purpose.

As a person, you likely have a job or you likely have responsibilities that you perform by yourself. You are autonomous in cases where you decide what you need to do. Autonomy refers to the ability to assign laws to yourself so that you follow them. This means that whenever you do a specific task, how you go about that task is a practice of assigning yourself laws.

Therefore, an autonomous vehicle refers to a vehicle that is capable of assigning itself laws without the need for human interference. For instance, if you were driving a vehicle that had no break, would you hit one person, or would you hit 10 people? This is a common morality question and it rightly has to deal with vehicles and whether it's a good idea for autonomy. Most people believe that if left to choose on its own, the vehicle would simply choose the single individual because of quantity. To those who see robots taking over in the future, obviously, the choice is 10 people. However, the machine, knowing that it might hit a person could choose an option that probably wasn't relevant to you at the time of me introducing this morality question. The machine would simply crash on purpose, which might damage the driver but is not likely going to kill them.

The morality question usually involves trains and not being able to make such an additional option in such a case. However, in the real world, that question is rarely going to actually exist. The ability to see an additional option and be able to take action on it without us assuming the choices it will have allows a vehicle to be autonomous. We, as humans, would commonly assume that there are only two options, but the vehicle would see additional options. Therefore, if the choice was left up to us, lives would be lost unnecessarily. This is the greatest point of autonomy for machines, allowing them to choose the best option without humans choosing that option for them.

**The Variables in Vehicles**

However, choices are based on variables and if we have an autonomous vehicle, that autonomous vehicle has to have variables. The first variable is the road itself, which is to say is it a straight road or is it an intersection or is it curved? All three types of Roads would lead to a different reaction from humans, which means that the shape of the road is ultimately the first variable that this autonomous vehicle has to deal with.

Just as it has to deal with the road, it also has to deal with the driving laws of the state or country or province that the car is in, which includes the speeding limit. This is so that the autonomous car follows human-made laws that have ensured human safety, which is the point of training a vehicle on the road. An autonomous vehicle will always be able to drive straight, it's literally just one variable that's being affected. It's only when we introduce the need for safety that extra variables are added it.

Therefore, along with knowing the speed limit, the autonomous vehicle now needs to be able to see. The vehicle needs to be able to see so that humans walking in front of the car will cause the car to stop. However, humans don't really see in the same way that computers see. There is a fantastic library called Open Vision that allows a computer to recognize elements that are important in the visual realm. The way that the computer would recognize that there are humans is if there is a bipedal shape crossing its vision. This would allow the car to react within the time that it took to recognize the bipedal shape.

The last variable that vehicles need in order to be autonomous is the weather, but this can actually be done with the internet, right? Actually, not really if you want it to be fully autonomous. There are places out in the world that don't have access to the internet, even in countries where internet seems to be practically everywhere. In such cases, you don't want an autonomous vehicle driving through the rain if it doesn't know that it's raining. Humans react differently in the rain because they need to be more careful as their vision is limited, which, by the way, the computer will have almost the same amount of limited vision with a few exceptions. For instance, humans cannot see in infrared whereas computers simply have a lens switch. Humans cannot have night vision, which is something that cameras can have. This allows the Open Vision system to see in environments that would normally be difficult for humans.

**A Combination of Brilliance**

Therefore, all in all, in order to create an autonomous vehicle, you have to have a combination of extremely brilliant inventions. The vehicle needs to be spatially aware so that means that proximity sensors need to be inside of the vehicle. The vehicle needs to be able to see humanoid shapes as it is driving along so as to not kill anyone. The vehicle needs to be able to understand the weather that it's driving in so that it can drive appropriately, which means that it needs to be able to have weather forecasting built-in. These are separate industries coming together to make it possible for a car to drive itself, which is one of the biggest feats of science ever to be created.

**Common Fears**

**Replacement of Transport Drivers**

The primary fear when it comes to autonomy in vehicles is the replacement of jobs. The problem with making advances in science and society, in general, is that there is always going to be a loser as a result. As of right now, a lot of people are worried that some who work as Uber drivers would be replaced by these machines since then Uber wouldn't have to pay for humans to drive around. This argument is similar to when Uber came out and taxi drivers were worried about being replaced.

A lot of people don't realize that a lot of people are very suspicious, which means that this new technology is not going to be widely accepted. It's going to have a few decades before any of us even trust it as much as we do the Windows operating system. Keep in mind, many of us still don't trust Windows operating system. Therefore, no, it's not going to replace Transport drivers, but it will supplement the availability of Transport drivers. I, for one, wouldn't mind being able to drive to locations for literal pennies. At least I know that an autonomous car is not going to mug me or rob me, which is not to say that the Uber driver will, but the likelihood is zero with the autonomous vehicle and unknown with the Uber driver. It would provide individuals who use Uber as a cheap option to go around town with an even cheaper option and thus the Uber driver would then become a premium experience. Therefore, when autonomous vehicles come out, you would actually see a rise in Uber payments because the human element would be appreciated more by the customers who want that human element.

## Unblamable Death

The secondary fear when it comes to autonomy and vehicles is who is going to be blamed for the death. As of right now, there have been a few companies that have made accidents that have resulted in the loss of lives and injuries in other cases. In those cases, the company that put the vehicle on the road has been sued and thus that is the end of the tale. The truth of the matter is that this is what would happen with autonomous vehicles.

The problem that seems to be bothering people is that if someone owns a vehicle that is autonomous by themselves that the person is responsible. However, if the vehicle is meant to be driven by itself then it is the manufacturer who is at fault because they are the one who accidentally left something out that resulted in the incident. Now, there is the highly likelihood that the company will shove the responsibility to the user as the user is supposed to pay attention and ensure that the vehicle is driving, but this would also make the vehicle pointless in many people's eyes. When people think of autonomous vehicles, they often depict what is shown in movies where the humans don't even bother learning how to drive. They simply call up their vehicle and go to another location.

However, companies do not want to assume risk in their vehicles and will likely treat autonomy the same way that companies treated cruise control. Cruise control allows you to maintain a certain speed, but you are still responsible for what occurs in and outside of that car as it is related to that car. For autonomy, companies will likely sell it as a feature much like cruise control that you will still be responsible for keeping an eye on the machine. This ensures that the driver of that vehicle is still responsible for any death that vehicle causes because it is a feature, not something to be relied on 100% of the time. It's a very clever way to get out of responsibility for when software doesn't work on the company's part.

**Google Map-esk Incidents**

The last fear, which is the less popular fear, is that the vehicle will act very similar to incidents that have happened in the past with mapping Services. I mentioned Google Map as it is one of the most popular methods to navigate around cities and countries. However, there have been moments where Google Maps is not entirely accurate. For instance, if you have a private community, Google Maps is not allowed to actually take photos of what is inside of that community. A lot of people who suddenly join a private community often find themselves having to give delivery drivers directions on how to navigate a neighborhood because Google Maps is not allowed in there.

However, Google Maps can actually lead to the wrong directions. For instance, Pokemon Go had an incident occur several times with its application because people simply wouldn't pay attention to the outside world as they did their exercise. This led to the injury and death of many people because the maps themselves were flat and didn't show that there was anything to worry about. This was, rightly so, the fault of the person who didn't look up from their phone, but the people still were outraged that this occurred. The people blame the company, when it was really the stupidity of the individual who didn't look up. This is the only fear that doesn't have an immediate solution to it and it's only because people get outraged when they are told they have to be responsible for it.

## Benefits of Autonomous Vehicles

### No More Drunk Driving

The benefit of having an autonomous vehicle is that you no longer rely on the human inside of it. Numerous deaths and accidents are caused by the humans inside of the vehicle not being fully aware. While the person inside of the vehicle might be drunk an autonomous vehicle would be able to navigate that person home by itself without putting anyone at risk at the same level of the individual who is drunk. This means that there would be a significant decrease in deaths and accidents caused by drunk driving.

### Optimizing Traffic Incidents

There are several traffic incidents that are often caused by humans not being aware of their surroundings. For instance, a person can only look so far outside a vehicle and many companies stupidly like to put things in front of their business that make it more difficult to drive. Instead of having the view from the front of the vehicle, which would be the safest way to view turning left or turning right, the human is usually in the center of the car. Small incidents like this could be avoidable with an autonomous vehicle because every inch of the car could then be loaded with sensors and cameras that would allow it to react to its environment.

## Software Knows All Laws

Furthering that point, a lot of accidents and tickets as well as jail time is caused from an ignorance in the law that the human is supposed to follow. For instance, everyone knows that jaywalking is technically illegal, yet it is rarely enforced. In fact, if it wasn't for jaywalking, cars that were being built for autonomy would not normally need to handle cases where they might hit a human.

## Cheaper Taxi/Government Taxi

I talked a lot about Uber and how the autonomous version of Uber would result in a cheaper tier of Uber drivers. However, that could also mean that the government could, instead of providing a bus for localized transportation in Big City, they could instead provide car driven Uber-like drivers. This would be much more helpful to individuals who need to get to work on time as they would be able to rely on the government service that drove them to work. This would affect a lot of people across the board, but it might also have some implications as a result.

**The Significantly Disabled Can Own Cars**

In the cases where a human is blind or in the case where a human is deaf or in the case where there is something physically wrong with the arms or limbs associated with driving, this vehicle could do it for them. A lot of the individuals, if you are disabled, usually have the right to drive stripped from them and this makes their life significantly harder if they need to go significant distances. Autonomous vehicles could, essentially, provide those disabled individuals with a way of owning a vehicle and normalizing their life after whatever they've experienced.

# How will it Impact traffic?

## No More Traffic Jams

This one will really only apply if almost all the vehicles on the road are autonomous vehicles. This is something that can be seen in many sci-fi films and novels where cars are able to control how the traffic flows. Given sufficient data and perhaps a network connection to the motor grid, you could technically create a system where no traffic ever has to stop. This is because the only reason why traffic stops is often due to the fact that humans need time to let other humans go the way they need to go. However, it is possible for cars to create a situation where no cars need to ever stop. It would be different than the way we experience traffic now because it would fully rely on automation. However, in the beginning steps of this automation, we would likely see a lot less traffic jams that were caused by accidents or incidents where the police pulled an individual over.

**Less Death**

As I had mentioned with the drunk driving and the ability to be a driving law book, there would be a lot less death due to traffic problems. This is primarily due to the fact that humans react slower than machines and so lives would be saved faster in emergency situations. Additionally, things like accidentally going before the light actually turns green would grind to a halt for most cars.

# Artificial Intelligence and the Job Market

**How many jobs will be replaced and why you should care?**

### All Jobs Will Be Replaced

Let's cut to the chase because every time that this topic comes up, it's really associated with the job that's being replaced. The problem with this topic is that people see the short-term result of what a specific technology will do. The truth of the matter is that all of the jobs in current existence will be replaced, given enough time. It is extremely easy to define and put into place machines for cooking hamburgers. The fact that we refrigerate hamburgers that are pre-made proves this point. Machines can already do most of the basic work that we need them to do.

It is not difficult to automate something, but what is difficult is choosing what needs to be automated. If you have ever talked to an accountant before, you will find that most businesses have different situations for their financial needs. It may be very easy to replace the hamburger maker or the cook in the kitchen with a robot, but it is much more difficult to replace the person who can assess the situation. The problem with artificial intelligence is that it doesn't have something that allows it to understand context.

Whenever you go to pay your bills, you may not pay your bills on time on purpose. In an automated system, the bills will be paid on time every time. However, you may need to wait a week to pay a specific bill because of a certain reason. Tons of people delay paying bills because of reasons and so one problem that robots have is understanding context, the reason why something is being done. This means that even though all jobs will eventually be replaced, the jobs that will be replaced last are the jobs that require context. You cannot automate the process of building a full-scale website, you can automate the design process, the building blocks, and many of the different elements of a full-scale website but that website changes based on the company needs.

## Even the Creative Jobs

This means that eventually even the creative jobs will be replaced once artificial intelligence machines can understand context. Here's the problem though; why does it matter? Why does it matter that jobs will be replaced? Jobs exist so as to continue our survival and basically to give us something to do until we die. It's not really a bad thing if all the mandatory jobs are replaced by robots because there is always going to be something else to do. Okay so you don't have to make hamburgers, but you can choose to create a shop of human-made food. That will become a specialty, shops that pride themselves on using only Human Services. Sure, you could automate a car fully, but a car is not going to speed down a runway at top speed to give you a drill thrill. Robots are designed to repeat repetitive tasks, things that you do for fun are things that only humans can do will exist. Only humans can make human made food or human made clothing, something that will be seen as the new fashionista style. The job market is always going to exist and there will always be something to do, you just have to have the right perspective.

# What do you need to know to implement A.I.?

## Easier Than Ever

The first thing that you need to know and understand is that you are standing on the shoulders of giants. A lot of people don't like to start out with this because they might think it's a little arrogant, but if you are just getting into artificial intelligence then you need to understand this. There has been a lot of work done in the past two decades regarding artificial intelligence, which means you are going to need to do a lot to get to where the frontier is at. That isn't to say that you can't do it within a reasonable time, it's just that you need to appreciate the complexity of this industry. Additionally, you need to understand that it has taken a lot of work to make things easy and while there are very easily implementable tools out on the market, understanding the core mechanics of how neural networks work is key to using these tools. The tools simply allow an individual to get the work they need done without having to deal with much hassle, understanding how those tools work is still something you're going to need.

## Algebra to Calculus

The second thing that you're going to need to know is a variety of mathematical skills depending on what you want your artificial intelligence to do. If you want your artificial intelligence to simply forecast the next week's stock prices, you'll mostly need to know statistics as well as maybe a little calculus. If, on the other hand, you want to utilize artificial intelligence to generate 3D pieces of Art then you may need some geospatial mathematics along with a little bit of discrete mathematics. There is a wide range of mathematical skills that may be required depending on what you want to do with it, but the reason why I specifically state algebra to calculus is because you will at least need to know algebra. Neural networks are designed with the understanding of the slope-intercept form as the most basic form of a neural node. It only gets much more complicated after that. Most of your learning will actually be solely mathematical and very little of it will be programming, but that is the third thing that you need to know.

## Programming

You will need to understand programming to the degree of the tool that you plan to use. If you are going to a website that allows you to use a neural network that was already built beforehand, you're likely not going to need much programming. If you plan to utilize a localized version of a neural network, you are probably going to need to know how to program and access the graphics processor unit Library that's compatible on your computer. A lot of people misunderstand this requirement because in the beginning they are thinking about DirectX 11 or 12 or maybe a Vulcan architecture, but these are graphical libraries. If you plan to create a localized neural network, you will need to know a significant bit about the hardware that you plan to use. This is because you can use the central processing unit or the graphical processing unit to do the job, but how you go about using it is definitely different.

These are pretty much the different things that you need to know in order to implement and create a neural network, which is the most of what people are after when they talk about artificial intelligence. You need to know the mathematics to create the neural network, you need to know the language needed to implement it, and finally, you need to know what way you plan to implement it.

**Which jobs will be replaced the soonest?**

**Repetitive Tasks Are the First to Go**

As I have mentioned several times at this point, the first jobs that are going to go are the ones that can be repeated. Flipping hamburgers, filing, writing checks, lifting things, stocking things, ensuring things are on shelves, and pretty much anything that requires a routine. That's almost all of the low-end jobs, the ones that teenagers and the elderly tend to find themselves at. These jobs will be the first to go because you don't need to pay wages to a robot and all you need to do is maintain the robot to extract the benefits. You will still need somebody in a managerial position to handle customers, but generally, all the basic jobs can be robot.

Now, it is important to understand that there will still be one person left to just be there. This is sort of like the individual that is there at the self-checkout. The individual is not really supposed to make sure that you are checked out and get all your groceries, they are there should anything go wrong. These jobs will become the new jobs that teenagers and elderly fit instead of the ones that require the person to check out. This means that Mom and Pop shops will probably still hire the person willing to look after the register during the business hours, but a company like Walmart or Target is likely going to hire one person to manage stocking robots.

There will also be an increase in need for Maintenance Technicians and Maintenance Engineers, to ensure that the robots are properly maintained.

## Jobs carried out via Rules Go Second

We've already begun seeing jobs that require rules begin to have their own version of replacements installed. For instance, as I mentioned before there is now a contract lawyer artificial intelligence that would essentially replace lawyers that focused specifically on contract work. These positions primarily follow rules and patterns, which means that even though it is significantly more difficult in routine than compared to stocking something on the shelf, it can still be automated given enough work.

## Consultancy Goes Third

The last type of job that will go is consultancy and the reason why I say this is because consultancy is a routine but contextual job. Sure, you could say that in consultancy all you are doing is judging what can be added or subtracted from a workload so that the company makes more money. This is something that a machine can currently do, but the problem comes in the form of contextual understanding. You see, any machine can go and create optimization methods for a business, but the business has to create that machine to fit that business. This means that the business itself is providing the contextual understanding the business needs in order to make an effective evaluation of what is needed to optimize the business. When a person comes in to consult for a business, they need to understand the business before they begin suggesting anything. This necessity for a contextual understanding is something that can't be quantified by a machine just yet and so this is why that will be the last type of job to go. However, ultimately, it will eventually go.

## Which jobs are least likely to be replaced?

### Inventors

The primary job that will not be replaced, I repeat it will not be replaced is an inventor. An inventor is an individual who thinks outside of the box. They look at the market and they look at what available tools exist before they begin generating ideas for what can exist if you combine those tools. The reason why an inventor will not be replaced is because almost all companies require an inventor in order to begin a company in the first place. They are the thing that drives the industry. They absorb more data than any current processor or processor within the next decade would be able to sustain and abstract into an invention. In other words, inventors don't have any rules beyond the rules of the universe. This means that you can't automate the job because there is nothing to automate.

## Frontier Science

The next type of job that is not likely going to see any form of automation is Frontier science and this is primarily due to the fact that scientists want to keep machines away from science. That isn't to say that there won't be a lot of science that these machines are capable of and it isn't to say that these machines won't be helping to march forward in the frontier science, but machines are not likely to be the entity to march forward in the frontier science. There's too much mistrust of machines, there is too much paranoia around the singularity, and if we hand over science to the machines then there will be nothing left for Humanity to do.

## Will Universal Basic Income fix the problem?

### Giving Everyone a Base Income

The idea of universal basic income is to give everyone a base income so that no one starves to death. This idea is not new and in fact, a lot of communist countries, as well as some socialist countries, believe in basic income for everyone in society. Due to the rapid replacement of jobs that might occur as a result of technology, many of the top leaders in technology have begun suggesting a universal basic income to offset the job loss. This would be provided on a global scale so that everyone could better their lives and it's a really good ideal but not a good idea.

Here is the concept in a nutshell because I have to describe more than what Universal basic income is as to why the leaders of Technology would believe in such an idea.. I mean bad idea. If everyone loses their job, no one has to suffer because people can still buy things if they have money. In order to ensure that they have this money, the richest people in the world donate so that everyone has a base income. This base income level would ensure that people could buy the bare necessities that they needed in order to live. This would not fix the problem of job loss, but it would significantly decrease the harmful impact that the job lost would have on the average individual because that individual would be able to buy food and similar items that would stimulate the economy.

## Companies Pass Cost to Customer

The problem is that the world doesn't work like that. You cannot have a society that was previously based off of meritocracy immediately converge into a community that shares everything, it just doesn't work. I'm not saying that the idea of a universal basic income is impossible, what I am saying is that when you spend centuries building a society towards always making more and not sharing, it becomes incredibly difficult to become a community that shares everything. Universal basic income would cause companies to pass the cost of that base income to the customer by making products more expensive. The problem is that you now are automating most of the jobs that currently exist, firing employees that would have made more than the universal basic income, and now you have an influx of people who receive your money to buy your products.

## Devalue the Currency

When you inflate the value of a currency, it will almost immediately depreciate in value. Let's say that we decided to put in the universal basic income and companies passed the cost on to the customer. You now have everyone at the same pay level if they don't have a job, However, money is limited. You may give them all a base pay grade, but the money is limited so unless you're going to print more money then you would have to take the money of the rich. If you take the money of the rich, the rich don't have any incentive to build new companies to become even richer if that just means you're going to take more of their money. On the opposite side, if you decided to print a lot of money you would devalue the currency. Printing money inflates how much money you have in society and this money is actually a representation of Exchange. Imagine that you had 10 tickets that you could trade in for a $300 guitar. Each of those tickets is worth $30. Now, printing money is the same as printing more tickets so let's say that you print more tickets and now you have 20 tickets for the same $300 guitar. Now, each ticket is worth about $15. Therefore, based on the example, you can pretty much see why printing more money makes money more worthless.

## Everything is Now the Same but Worse

So now that you have seen the concept and you see what results from it in a very, extremely simplistic example, you can see why this is a bad idea. Sure, in the first month, maybe, the base income becomes useful but every month after that you have companies passing the cost on to the customer and then you have the worst, which is that the money that pays the cost lowers in value. This causes everything to become more expensive and the base income value is now pointless because everything you could have bought with the base income to survive is now more expensive. Therefore, everything is mostly the same but worse because now what you're getting paid is worth less than it was originally. This is why universal basic income simply doesn't work and why it has failed every society that has tried to use it.

# Don't Be Scared; Use It

## A.I. is just a Number

You need to understand that artificial intelligence is really just a probability matrix of what should be chosen and what should not be chosen. A lot of people think that when this reality comes into fruition that robots will kill everyone and rule the world, but the truth of the matter is that even the most complex robots that are out there today are not thinking for themselves. The rules that are in place inside of their systems are designed by programmers, which is to say that they are rules that are predictable.

All of this is just really scaled-up mathematics that has been incorporated with machines so that they are able to do the things that we need them to do. Knowing that this is something that the average person can do, it shouldn't be something that you are scared of. You should realize that almost all of the robotics makers in the world don't really have a specific intent to create robots that are designed to eliminate humans. People can't become rich off of these, people will immediately crucify any attempt to do this, and it will have a global outrage at the first person to attempt to do this. Essentially, it is one of those actions that the entirety of the world would condemn and then we would have a lot of laws that make it difficult for another person to do it.

On top of this, it is extremely difficult to create something like the Terminator. You have to realize that these robots can only work with soft bodies and we are just now exploring how we can create those soft bodies. Machines like the Terminator are significantly heavier, so something like a turret inside of a robot is not feasible because of physics. Rockets are not feasible because of physics. Essentially, if you actually look at how the Terminator is designed, you will quickly find that the Terminator that everyone is afraid of is a physical impossibility as a weapon. As an artificial intelligence that's stronger than most men, that is possible. However, having an arsenal of weaponry contained inside of the body is an impossibility if that robot is the size of a person. People are precarious designs at best and not really good at maintaining a center of gravity. In fact, this is why it is so tricky to create a human-like machine; the physics needed for the humanoid anatomy is extremely difficult to master. You might see something that comes out like MechWarrior that might have that intelligence, but you are not likely going to see a robot that looks like a human that suddenly pops out a rocket. This is reality and reality is based on numbers, if you look at the numbers it is extremely difficult to come to the fear that robots will take over in the way that Terminator took over. There's just simply so much that's wrong with the core principle that wouldn't work in our world. Don't let the fear-mongering around a robot apocalypse to decide how you want to utilize machine learning and artificial intelligence in your daily life. There are thousands of ways the world can end, the robot apocalypse is only one of them and it's not really possible.

## Decisions Made Quick

It doesn't matter whether you are working in the health industry, stock market, real estate industry, or any other industry required in making decisions. Due to the fact that machines can go through decisions must faster than humans, they are able to make a much bigger change in society. They are able to quickly to decide how you can make more profit, how people can be safer, and generally decide things that benefit Humanity in a way that's faster than Humanity can decide for itself. Let's walk through an example.

Let's say that a car will fall if it falls off of one side of the building. However, once it starts rolling, you cannot stop the car. A human would have a few seconds where the human would not decide anything, and it would take them a few seconds to understand everything that needed to be decided. A computer, on the other hand, would be able to look at the scene and decide what it needed to do within nanoseconds of understanding the rules. Machines can decide things much faster than humans and can result in better improvements at a faster rate than humans can do themselves. Don't let giant companies decide the future if you can put your foot forward with barely any barriers on what you want to do.

## Less Repetitive Work For All

No one really likes doing repetitive work unless it is something that is not normally seen as repetitive work but a type of hobby. For instance, there are quite a number of people that love fishing but fishing is actually a profession done by a few people otherwise we would not have Red Lobster. It's not really seen as a job until the specifics are talked about.

Artificial intelligence allows us to remove the annoying jobs that most of everyone don't want to do. No one really wants to focus on driving for 5 hours at a time, no one really wants to pick vegetables from 100 acres, no one really wants to ensure the cleanliness of the sewers themselves, and these are all things that can be handled by artificial intelligence. Don't let the fear that no one will be able to find work stop you from innovating and allowing people to take on careers they prefer more. You can decide whether there will be a lack of jobs or if the jobs that are available are jobs people love doing.